S. M. Wells

The Electropathic Guide

Prepared with Particular Reference to Home Practice

S. M. Wells

The Electropathic Guide
Prepared with Particular Reference to Home Practice

ISBN/EAN: 9783337812348

Printed in Europe, USA, Canada, Australia, Japan

Cover: Foto ©berggeist007 / pixelio.de

More available books at **www.hansebooks.com**

THE

ELECTROPATHIC GUIDE:

PREPARED WITH

PARTICULAR REFERENCE TO

HOME PRACTICE;

CONTAINING

HINTS ON THE CARE OF THE SICK, THE TREATMENT OF
DISEASE, AND THE USE OF ELECTRICITY;

WITH

Full Directions for Treating over 100 Diseases.

BY S. M. WELLS,
MEDICAL ELECTRICIAN.

FIFTH EDITION.

LA CROSSE, WISCONSIN.
1879.

PREFACE.

In offering this little book to the public, it is but just to say that it was originally designed for our pupils, and those who had attended our lectures, as it was through the earnest and repeated solicitations of such that the work was undertaken, but before it was completed, so many calls were received for it from other sources that a few changes and additions were made, with the hope it would in some measure meet these demands also.

These changes were made with particular reference to those who have had little experience in the use of Electricity, or in the treatment of diseases, for though it is no easy or unimportant task to prepare a work in every way suited to the wants of the inexperienced and uninformed, so many of this class are at the present time resorting to the use of Electricity, that we thought it was better for them to have even a little light than to work entirely in the dark. Then they have our sympathy, and we would benefit them if we could, as most of them have tried the various remedies employed in the treatment of diseases without obtaining the desired relief, and in their great suffering turn to Electricity as the last resort and only hope.

For the encouragement of such we would say, work cautiously, and you not only need not stumble, but may rejoice over happy results, for diseases frequently yield to Electrical treatment after various other remedies have failed to reach them.

As many of those for whom the work was designed have not been favored with a medical education, and therefore are not familiar with medical terms, we have avoided all technicalities, and endeavored to give a plain statement of facts in as explicit and concise a manner as possible, and in a language so plain and simple that all who read may understand. We have, also, in consideration of the inexperience of these, omitted some diseases which may be treated with success by experienced and skillful Electricians.

With the hope that it will prove useful to others, as well as to those for whose benefit it has been prepared, and bring health and comfort to many a *suffering* one, the work is therefore submitted to the public.

CONTENTS.

CHAPTER I.

ELECTRICAL INSTRUMENTS AND APPARATUS FOR MEDICAL PURPOSES.

CHAPTER II.

ELECTRICAL TREATMENT.

CHAPTER III.

METHOD OF APPLYING ELECTRICITY.

CHAPTER IV.

ATMOSPHERIC ELECTRICITY.

CHAPTER V.

TREATMENT.

CONTENTS.

CHAPTER I.

As the laws which govern Electricity, and the principles upon which it should be applied to the treatment of disease, are universal, the instructions given in this work may be followed, (so far as the operator has the facilities for doing so,) whatever Electrical Machine or battery he employs, if it is reliable, and if the positive and negative poles of these batteries correspond with those of the battery here described, as most of them do, though there are exceptions.

A description of, with directions for using Wells' Double Current Electropathic Instrument, which we have been using with excellent success for several years, will, I think, enable one to operate almost any battery in common use.

This instrument is very powerful for its size, yet so simple in its structure that its action may be easily understood, and so arranged that it does not readily get out of repair. No acid is used in operating it; the solution, being one of Sulphate of Copper, (or blue vitriol,) if it is spilt on the clothes or carpet, no harm is done.

It possesses decided advantages over the single-current instruments, as both the currents are combined in one instrument, and can be brought to bear on the directors by merely moving a switch, or lever, without disconnecting the conducting wires.

The primary current possesses sufficient power to discuss all ordinary tumors, and to remove other extraneous growths. The secondary current is of sufficiently high intensity for all medical purposes. This instrument is so constructed, that it can be operated with the box shut, thereby preventing the noise of the vibrating armature to escape, and also keeping the instrument free from dust.

Fig. 1.

PROF. WELLS' BATTERY.

DIRECTIONS:

1. Connect the wires from the battery screws to the screws in the instrument, as represented in the cut.

2. The solution to be used in this battery is one of Sulphate of Copper. (blue vitriol,) con-

taining about two ounces of blue vitriol to a quart of water, as it takes about this quantity of water to fill the copper cup, which should be kept nearly full when in use. If a *very* light current is required, the solution can be reduced by adding more water.

3. The zinc cup or plates become coated in the battery, so that it is necessary to clean them whenever they become thickly furred. This should be done by scraping them with a knife, (instead of washing or brushing them,) so as to expose again the bright surface of the zinc.

4. The bundle of wires in the helix is to regulate the current. It can be removed or introduced at pleasure, thus regulating the strength of the current. When entirely out, the current is very light; as you insert it the strength is increased, and when it is nearly in, the instrument is in its full strength. This is called the regulator.

5. To get the primary current, move the switch, which connects with the screw-post, on the brass knob marked P.

To get the secondary current, move it on the knob marked S.

6. If the Electro-Magnetic apparatus will not operate, see first if any spark is perceptible on

rubbing the extremities of the wires from the battery together. If not, the battery is in fault. This may be owing to a sediment of copper in the bottom of the battery making a connection between the zinc and copper, or to the zinc being somewhere in metallic contact with the copper; or, it may be owing to the foulness of the zinc plates, or to the weakness of the solution, which, in that case, will have lost its color. If the fault is not in the battery, it must be in the vibrating armature, which is placed under the arch of brass. This must be adjusted so that the iron hammer is about the sixteenth of an inch from the face of the magnet; then screw the spiral spring down so as to touch the flat spring and tighten the screw by the lower nut, and there will be no difficulty in the instrument's operating.

As there are many local and internal diseases which cannot be reached without the use of in-struments, a case of instruments accompanies each of Wells' Machines, (unless otherwise or-dered.) These instruments are got up very neatly and substantially, and are so simple in their structure as to require but little skill in using them; while they are of great practical use, as there are many internal diseases which cannot be reached without them.

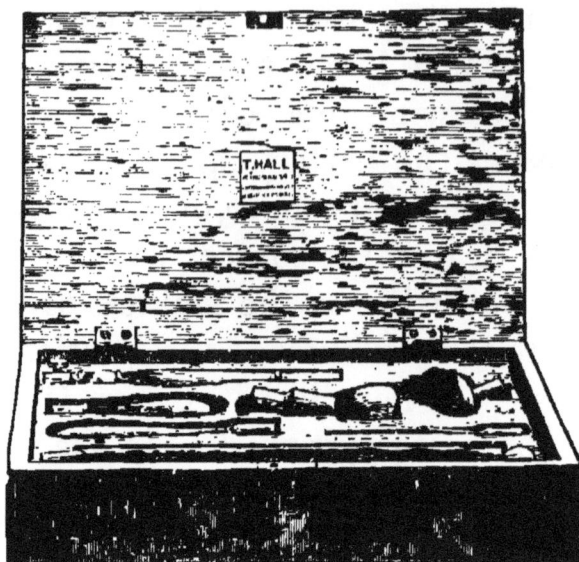

Fig. 2.

CASE OF INSTRUMENTS.

This case of instruments contains an Insulated Handle and Sponge, an Ear, Eye, Throat, Tongue, Womb, Vagina and Rectum Director, silver plated and glass. One instrument serves for the last two, (the Vagina and Rectum,) and is the largest silver instrument in the case. The Womb director, (which is the long glass instrument,) is to be attached to the conducting cord by means of a thumb-screw. All the other directors are fitted to the Insulated Handle, and can be put on and taken off at pleasure. As but one of them is used at a time, they need only one handle, which makes the set much cheaper than if each had a handle attached.

The flat instrument is for the tongue; the small pointed one for the ear, and other delicate purposes; and the long one, with the ball on the end, for the throat and fauces.

A bit of moist sponge, or cotton flannel, should be fastened on the end of the ear and throat instrument when in use. The tongue instrument may also be covered with cotton flannel, or soft linen, if the tongue is sore, but is usually used without covering, as is also the vagina and rectum instrument. When in use the eye cup should be filled with *pure, soft*, tepid water.

When the operator is treating a case which requires him to hold both poles, he should use the Insulated Handle and sponge in one hand, instead of both electrodes, as that will insulate him from the current.

It must be remembered that the sponges should always be wet when in use, and thoroughly cleansed after use.

CHAPTER II.

The general rule for treating diseases with Electricity is, that inflammations, fevers, bruises, sprains, expanded muscles, swellings and extraneous growths should be treated with the positive pole; and debility, chilliness, inaction, tendency to decomposition, and contracted muscles with the negative pole. In treating the first class of diseases, as far as possible, arrange the poles so that the positive will be above the negative. In diseases which require treatment with the negative pole, this rule cannot be observed.

Example—In inflammation of the eye, the negative is held in the hand, and the positive at the eye; while in amaurosis—or paralysis of the optic nerve—the negative is held at the eye, and the positive on the spine.

PRIMARY AND SECONDARY CURRENT.

The primary current is used more particularly
(18)

in treating unhealthy diseases, and extraneous growths, such as scrofula, cancers, ulcers, swellings, opacity of the eye, granulated eyelids, enlarged tonsils, enlarged joints, etc. And in these cases it should be used with the regulator out, or nearly so, and with the full strength of the solution. This current may also be employed in any case where the patient is too sensitive to bear the vibration of the secondary current; though, in most cases of this kind, it should be used with the regulator partly, or entirely inserted in the helix. If the current should then be too strong, (which is sometimes the case *when used on the head,*) the solution may be reduced by adding water.

The secondary current is better adapted to nervous affections, and inflammatory diseases, and is employed to arouse and give tone to the system.

DIAGNOSIS.

If, at any time, the operator is at a loss to know which pole to apply in giving treatment, he can be greatly assisted in making a correct decision, by testing the sensibility of the diseased parts to the current; for it is a remarkable fact connected with the treatment by Electricity that it detects not only the organs, but the very portions of the organs where the greatest amount of irritation exists,—a knowledge which a pressure of the hand often fails to elicit.

The following is the best method for making this examination of the internal organs.

Attach the sponge, which belongs in the case, to the Insulated Handle, and connect them with the positive pole. Moisten the sponge, and the surface of the body under examination, then hold the sponge on the spine, opposite, or a little above the affected part, while the fingers of the hand in which the negative is held are passed with a firm, though not disagreeably hard, pressure, over the diseased organ. The portion where the inflammation or irritation exists, will be found to be more sensitive to the current than when in a normal condition. If not fully satisfied in reference to the location of disease, a more general examination may be made. Wet the neck thoroughly, and hold the positive on the upper part of the spine, while you examine the throat with the negative; then lower the positive two or three inches, and examine the trachea and upper part of the lungs, and thus pass down over the entire body, keeping the positive on the spine, a little above the negative, in front.

To examine the spine, place the negative at the base—by seating the patient on it, or otherwise, —and examine each vertebra, by pressing the fingers firmly on it. Diagnosis of the head should be left to experienced Electricians.

The negative electrode, with a small sponge around it, may be employed instead of the fingers, in making the examination, though this method is not so reliable.

The primary current is seldom found to be of sufficient intensity for diagnosing. As an unnatural sensitiveness of the parts to the current indicates treatment with the positive pole, so a want of sensitiveness indicates treatment with the negative pole.

Note.—The operator should hold the *pos.* in his hand while examining the spine.

CHAPTER III.

A general or local application may be made, as the case requires, or both, if necessary. Local diseases cannot be readily cured without a direct application to the diseased part, and there are many internal diseases which can be reached only by the use of instruments, as, in order to effect a cure, the current must be conveyed to, or directly through the diseased organ. This can be done without pain, and in most cases, without even unpleasant sensations attending the application.

In giving treatment, the operator should be supplied with two good bathing sponges to wrap around the electrodes, or poles, of a proper size and shape to entirely envelop the metal, in order that it may not come in contact with the patient, and a basin of tepid or cold water, whichever is most agreeable to the patient, in which

(22)

to wet and rinse the sponges during treatment. Wet sponges are good conductors of Electricity, and prevent those unpleasant sensations which are so disagreeable to nervous, sensitive patients. They also remove the impurities from the skin.

GENERAL TREATMENT.

In a general application—or, in what we shall call in the following pages, General Treatment— the entire person, *except the head,* is sponged with one pole, or electrode, while the other is either at the feet, the base of the spine, or in the mouth, as the case may require. It is, however, seldom advisable to give General Treatment with the *positive* at the feet.

Examples.—For a fever, put the feet in warm water, and place the negative pole, attached to the Insulated Handle and sponge, in the water with the feet, then sponge the entire person with the positive. If there is severe pain and heat in the head, it may be thoroughly wet, and also treated with a *very light.* current. The Insulated Handle is used in this case, that the cord may be kept out of the water, as it will last much longer if kept dry.

GENERAL DEBILITY, OR TONIC TREATMENT.

Wrap a wet sponge around the positive pole, and seat your patient upon it, or, if in bed, place it at the base of the spine, then sponge the entire person (head excepted) with the negative. If, in cases of debility, there is also a tendency to sore throat, or flow of blood to the head, with cold extremities, the positive pole may be attached to the throat or tongue instrument, and first placed in the mouth and held there by the patient, while the upper half of the body is sponged with the negative; then placed at the base of the spine, as before, while the lower half of the body is sponged with the negative.

In giving General Treatment, do not wet the entire person at once, as that would be apt to produce chilliness, especially in feeble patients; but commence by sponging the neck and shoulders, then wipe dry and rub with the hand until warm, then take an arm and treat in the same manner, and so on, until the entire person has been treated. If the patient is inclined to chilliness, keep him covered with a blanket and place a bottle of warm water at his feet.

LOCAL TREATMENT.

In Local Treatment, the poles must be so ar
ranged that the Electricity, which flows quite
directly from one to the other, will pass through,
or to the diseased part under treatment.

Example—In inflammation of the stomach,
pass the positive over the stomach, while the
negative is held on the spine opposite, and a little
below the stomach. In this case the Insulated
Handle and sponge should be held on the spine,
instead of the electrode.

STRENGTH OF CURRENT TO BE USED, AND TIME OC-
CUPIED IN GIVING TREATMENT.

No exact rules can be given for the length of
time or strength of current to be employed.
Both must be varied to meet the condition of the
patient and the parts under treatment, as some
persons are naturally more sensitive to the current
than others, and, in all persons, certain parts of
the body are more sensitive to it than others; besides, disease either increases or diminishes the
sensibility. Draw the regulator out so as to commence with a light current, and increase it gradually. Except in cases where there is inflammation or soreness of the parts, the current should

not be of sufficient strength to be painful or particularly disagreeable to the patient. These cases are often sore to the touch, and must, therefore, be treated gently, though with a good current. In treating children the operator must be his own judge. A light current, of course, must be used, and a comparatively short treatment given.

The current must also be varied on different parts of the body according to the sensitiveness. The head is extremely sensitive, and should be treated with great caution. It should be thoroughly wet when treated, and seldom requires treatment with the negative pole. Many persons cannot bear the secondary current at all on the head, and not even the full strength of the primary current. The lungs, next to the head, are sensitive, and require a light current. The mouth, also, is quite sensitive, though much more so in some persons than in others, while the spine, stomach, liver and bowels will bear a strong current, unless irritated or inflamed.

The time occupied in giving General Treatment may vary from five to thirty minutes, and for local treatment, from five to twenty minutes.

TREATMENT, HOW OFTEN REPEATED.

In severe, acute attacks, the treatment may be

repeated two or three times during twenty-four hours, if necessary. In chronic cases, a daily treatment for a week, or ten days, and then from three to five treatments a week, will be sufficient. Obstinate cases of long standing frequently require a long course of treatment to effect a cure, and the symptoms are often aggravated, at first, by the treatment, though only for a short time. It is not best to get easily discouraged in cases of this kind, for, not unfrequently, those which seem to be hopeless are cured; or, if not, they are greatly relieved. Should the symptoms be much aggravated, omit treatment for a few days.

USE OF INSTRUMENTS.

When treating with one pole in the mouth, it is always to be understood that either the throat or tongue instrument is to be employed.

When any instrument is used in giving treatment, it is always advisable to place the instrument before applying the other electrode. Commence with the regulator out, or nearly so, and increase the current by inserting it gradually into the helix.

This is a good rule to be observed in all cases, as it will prevent severe shocks, not only to the patient, but also to the operator.

The Vagina and Rectum Director should be well lubricated with linseed or olive oil when in use.

The knob at the end of the Womb Insulator can be unscrewed and taken off, and the rod inside removed, if necessary, in order to cleanse it.

Great care should be taken to cleanse the instruments and sponges after use.

If a patient wishes to treat himself, in a case where it is not convenient to use the Insulated Handle, India rubber gloves may be worn to insulate the hands.

CHAPTER IV.

We would recommend Hall's Glass Castors, for Insulating Bedsteads, etc., to invalids who are confined to their beds by chronic diseases, and to persons who suffer from languor and exhaustion on rising in the morning.

Sleep should be refreshing. It should be "tired nature's sweet restorer," and, if it is not, there must be a cause. This cause is, without doubt, in very many cases, the constant escaping of the electrical currents from the body to the ground, which takes place more rapidly while in a recumbent position, and during sleep, than at other times. These castors prevent the currents of Electricity from passing off from the body to the ground, by insulating the bedstead. They are made of pressed glass, about three and one-half inches in diameter, and one and a half inches thick, with a cavity to allow the feet of

the bedstead to rest in. By being smooth on
the bottom, they will slide on the carpet much
better than the metallic castors. They are also
much used as a protection from lightning. In
order to insulate the bedstead it must be removed
from the wall, so that it will not come in contact with anything but the castors.

We would also advise sleeping with the head
toward the north, (in this hemisphere,) and the
feet toward the south, in order that the strong
currents of Electricity, which are constantly flow-
ing from the poles towards the equator, may
pass in the same direction as those in the body,
which flow from the brain downward and out-
ward, as the action of these strong opposite currents has a tendency to break up the equilibrium
of the natural currents in the body, thus produc-
ing wakefulness, restlessness, and even great
nervous derangement in persons of delicate, sensitive organizations, and in those where the system is already diseased.

That the human system is decidedly affected
by atmospheric Electricity, we have unmistaka-
ble proofs. Why is it that periodical headaches
return at the same hour day after day? Why
do certain diseases prevail during certain condi-
tions of the atmosphere? And why is it that

the rheumatic patient so unmistakably foretells
the approaching storm, while one that is suf-
ering from severe bronchial irritation becomes
aware of a "change in the wind" even at mid-
night, while enveloped in the warm coverings
of a comfortable bed in his own closed apartment? Says Dr. Garratt: "The human organism is decidedly affected by atmospheric Electricity, from the slighter changes in the electric
state of the air, as well as by *bolts of lightning*.
Healthy individuals even feel the exhilaration
of a serene and positive atmosphere, as also an
increased heaviness and oppression at the opposite state of the air. If the weather be stormy,
and the air is surcharged either positively or neg-
atively, or is suddenly changed from one to the
other state, then do we find the neuralgic, the
rheumatic, and all invalids feel pains and de-
pressions. When the lower strata of the air,
which is near the surface of the earth, obtains
also in the higher strata for awhile, as before a
storm, and sometimes just after, then it is that
the rheumatisms ache and the neuralgias give
their ugly twinges; the frail feel a peculiar fatigue and are irritable, or are perhaps drowsy.
In the extremely sick, the dyspnœa of emphy-
sema and of heart diseases are worse; complica-

ted chronic rheumatism is awakened; paroxysms of fever anticipate their accustomed hour; in severe acute cases the symptoms become doubly alarming; while in fatal cases, death arrives earlier in unsettled and stormy weather than would have been had the atmosphere been serene. Now, all these coincidences cannot be accidental; and as the operation of natural laws is as unerring as that of the most exquisite machinery, due investigation and inquiry only are requisite to lay bare this whole mystery."

Daily observations and experiments were made in and about Paris during the cholera there in 1849, which show a striking coincidence between the malignance of the disease and the amount of atmospheric electricity. These observations and experiments were made by M. Andraud, and in a letter to the President of the French Academy, dated June 10th, 1859, he says: "The Machine I have used for my daily observations is rather powerful. In ordinary weather it gives, after two or three turns of the wheel, brilliant sparks of five or six centimetres. I have noticed that since the invasion of the epidemic, I have not been able to produce, on any one occasion, the same effect. During the months of April and May, the sparks, obtained with great

trouble, have never exceeded two or three centimetres, and their variations accorded very
nearly with the statistic variations of the cholera. This was already for me a strong presump-
tion that I was on the trace of the important
fact I was endeavoring to find. Nevertheless, I
was not yet convinced, because, one might at-
tribute the fact to the moisture of the atmosphere or to the irregularities of the electric machine. Thus I waited with impatience the arri-
val of fine weather, with heat, to continue my
observations with more certainty. At last fine
weather came, and to my astonishment, the machine, though often consulted, was far from
showing, as it ought, an augmentation of electricity, but gave signs less and less sensible, to
such a degree, that during the 4th, 5th, and 6th
of June, it was impossible to obtain anything
but slight cracklings without sparks. On the
7th, the machine remained quite dumb. This
new decrease of the electric fluid has perfectly
accorded, as it is only too well known, with the
renewed violence of the cholera; for my part, I
was not more alarmed than astonished; my con-
viction was complete. I saw only the conse-
quence of the fact already supposed. It may be
imagined with what anxiety, in these moments

2*

of the crisis, I consulted the machine, the sad
and faithful interpreter of a great calamity. At
last, on the morning of the 8th, some feeble
sparks reappeared, and from hour to hour, electric intensity increased. I felt with joy the
vivifying fluid was returning in the atmosphere.
Towards evening, a storm announced at Paris
that electricity had re-entered its domain; to
my eyes, it was the cholera disappearing with
the cause which produced it. The next
day, (Saturday, the 9th,) I continued my observations; the machine then, at the least touch,
rendered with facility the most lively sparks."
Now, it is stated that in the six days following
the 8th of June, the mortality in Paris fell reg-
ularly from 667 to 355.

Dr. Pallas, chief physician of the French ar-
my in Africa, asserts that the greater number
of diseases, particularly the "neuroses," are due
to the influence of widely deviating electricity;
the principal sources of which are thunder clouds
and marshy soils. By their effects on the human
organism, and their geographical constitution,
marshes present the closest analogy to the gal-
vanic battery, and observation proves that diseases developed by marshy emanations, are, at first,
of a nervous nature; hence, one of the methods

which will be found most efficient in preventing intermittent fevers and neuroses, must be the *electrical insulation* of chairs, beds and tables from the earth by glass supporters.

Sir James Murray, M. D., after extensive observations and experiments, says: "I consider that the *exiting cause* of epidemics which is called *malaria*, is not *bad air* at all, as the name implies, but the result of disturbed electricity; that marsh miasms, gases, or effluvia of vegeto-animal matters, or putrid emanations, are not, as is commonly supposed, the exciting causes of agues or diseases called malarious. But I consider these noxious emanations are disturbed *electro-galvanic* currents and electric communications—sometimes positive, sometimes negative—causing a want of electrical equilibrium in human bodies, etc. I consider that men's bodies between the atmosphere and the earth, represent the chain of a Leyden Jar, or of an Electric Machine, conducting negative electricity from the outside of the jar to the ground, or supplying positive electricity from the earth to the rubber. Were the surface of the floor well *insulated*, the chain could not readily give or receive the currents which otherwise pass through it."

It had long since been proposed by Dr. Priestley to electrify a great number of patients at once by placing them in a chamber raised upon glass feet.

Mr. Ellis recommended, in 1831, that persons seized with cholera should receive their remedial treatment in beds placed upon glass bottles, and be supplied with their remedies in glass vessels. All these ingenious suggestions were proposed for the use of persons already diseased; but says Dr. Priestley: "The above gentleman has suggested means of *cure;* I recommend measures of *prevention.* Their propositions were never carried into effect; whereas, my insulated houses were *tried,* and saved the inmates from diseases in places where laborers, previously unprotected, fell by dozens in fainting and fevers, for want of sufficient electricity to sustain the natural balance. Persons insulated by a very bad conductor, such as a floor of cold asphalt, and by *clean, dry flannel,* or insulators, cannot readily communicate electricity to the earth, nor receive electricity from it, *if the air of the apartment be dry* where they sleep, and free from filth and moisture.

CHAPTER V.

NOTE.—*Pos.*, denotes the Positive Pole; *Neg.*, the Negative Pole; *P. C.*, the Primary Current; *S. C.*, Secondary Current.

ABSCESSES.

To prevent suppuration, apply the *pos.* to the swelling, and the *neg.*, when admissible, either at the hands or feet, according to the location of the abscess; if below the shoulders, at the feet; otherwise, at the hands, putting them in water, with the *neg. P. C.* (See *Tumors.*)

AGUE IN THE BREAST.

Apply the *pos.* to the disease, while the *neg.* is held on the spine a little below it, or in the patient's hand, on the affected side. Treat from ten to thirty minutes, as the case may require, and repeat the treatment in a few hours, if necessary. If the treatment is repeated, or if it is

(37)

continued more than ten minutes, the *neg.* should be held on the back, only a portion of the time. *P. C.* Give treatment as in fever, if necessary.

ANUS, PROLAPSUS OR FALLING OF.

First, the fallen parts should be replaced. Then connect the Rectum Director with the *pos.* pole, and after oiling well with linseed or olive oil, introduce it carefully into the rectum, holding it there while the *neg.* is passed up and down the lower part of the spine for ten or fifteen minutes, *S. C.* The patient should keep a horizontal position for an hour or two, at least, after treatment. For this reason, a favorable time for treatment is just before retiring for the night.

ANUS, INDURATION OF.

Wrap a small soft sponge around the *pos.* pole and press it against the parts, while the *neg.* is passed up and down the spine for ten or fifteen minutes, repeating the treatment four or five times a week until cured.

An occasional treatment may be given with the *neg.* at the feet instead of being applied to the spine. *P. C.* used in both treatments.

APPARENT DEATH.

First, place the *pos.* wrapped in a wet sponge on the back of the neck and apply the *neg.* for three or five minutes, over the lungs, commencing with the regulator *nearly out* and introducing it gradually until you have a strong current unless the patient begins to show signs of life; in that case, have the current only moderately strong. Also apply the *neg.* to the arms for a moment. Then place the *pos.* at the base of the spine and sponge the entire person with the *neg.* using a strong current. If the operator can rub with the dry hand, instead of the sponge, while the *neg.* is held in the opposite hand, during a part of the treatment, it is desirable *S. C.*

Warm flannels, and bottles of warm water should be used about the patient, at the same time great care should be taken to have fresh air in the apartment.

A minister in Wisconsin who was supposed to have died from heart disease was restored by treatment similar to the above, and is still preaching, though this occurred several years since.

APPETITE, LOSS OF.

The Tongue Director, (which is the flat instrument,) should be connected with the *pos.* pole, and held on the tongue as far back as practicable, while the *neg.* is passed over the stomach, liver and bowels. Commence with a light current, and increase gradually as the patient becomes accustomed to it. Be careful to always place the instrument on the tongue before making contact with the other electrode, *S. C.*

Give plain, spare diet. *Entire* abstinence from food is sometimes the best treatment that can be given.

APOPLEXY.

When accompanied by a hard, full pulse and flushed countenance, place the *neg.* at the feet, and both in warm water, if possible, (as in General Treatment,) and commence by applying the *pos.* to the spine, from the base of the brain down, beginning with a light *S. C.* and increasing the strength as you proceed in the treatment. After treating in this manner for ten minutes, wet the patient's head thoroughly, turn the switch on to the primary, draw the regulator nearly out, and apply the *pos.* gently to the head for three or four minutes. The treatment may be repeated in ten or fifteen minutes if the symptoms are not decidedly improved.

Be cautious that no mistake is made in arranging the poles, as that might prove disastrous to the patient. Keep the patient as quiet as possible.

ASTHMA.

Apply the *neg.* over the throat and lungs, bringing it as far down as the diaphragm, while the *pos.* is held on the spine, and kept a little above the *neg. S. C.* If the patient is debilitated, which is usually the case, give tonic treatment. The paroxysms may be relieved, but the time to cure is during the intervals, giving from four to seven treatments a week until there is an improvement in the case, then less frequently.

WEAK BACK.

Seat the patient on the *pos.* and sponge the back with the *neg. S. C.*

BRUISES AND SPRAINS.

Apply the *pos.* to the injured parts and the *neg.* at some point below—at the hands, feet, or base of the spine. In the case of a sprained ankle, place the *pos.* in water with the injured foot.

having the injured part entirely submerged in
the water, and the *neg.* under the other foot.   In
a severe case, a treatment of twenty or thirty
minutes may be given, and repeated two or
three times during twenty-four hours. A sprained wrist should be treated by putting the hand
in water in the same manner.   In an internal
injury, it may be necessary to hold the *neg.* opposite instead of below the *pos. S. C.*

BOWELS, CONSTIPATION OF.

The *pos.* should be held on the tongue while
the stomach, liver and bowels are sponged with
the *neg.*, then seat the patient on the *neg.* and
sponge with *pos., S. C.* Avoid cathartics, use
enemas, if necessary, eating slippery-elm freely
during the day, and a table-spoon full of wheat
bran before each meal, will aid in very obstinate
cases.   The bran may be eaten in a little milk
or sweetened cream.

BOWELS, INFLAMMATION OF.

Place the feet in warm water, with the *neg.*
in the water, and sponge the bowels thoroughly
with the *pos.* The entire person should be
sponged with the *pos.*, at the same time, *S. C*

Compresses of hot water on the bowels (covered with hot flannels,) is an excellent auxiliary in the treatment. The bowels should be kept open by the frequent use of enemas of warm water.

BOWELS, LOOSENESS OF.

Place the *pos.* at the base of the spine and apply the *neg.* with the sponge to the back, bowels, stomach and liver, *S. C.*

BLADDER, INFLAMMATION OF.

Hold the *pos.* over the diseased organ while the *neg.* is on the lower portion of the spine on or at the feet, *S. C.*

BLADDER, PARALYSIS OF.

Place the *pos.* at the base of the spine, then hold the *neg.* over the diseased organ, *S. C.* and *P. C.* alternated.

BRAIN.

Diseases of the brain require skill and experience. *A very light current should always be used about the head.*

BURNS AND SCALDS.

Should be treated with the *pos.* pole. Care should be taken not to rub the sore. Wet a soft, thin cloth, and lay it carefully over the sore; 'then hold the sponge gently on it. The best dressing for it is a thick lather of pure Castile soap, put on immediately and carefully with a clean lather brush, so as to entirely exclude the air, renewing it frequently until healed. Indeed, this is all the treatment necessary in most cases.

BRONCHITIS.

First, the *pos.* connected with the Throat Director, should be held in the mouth while the *neg.* is applied to the upper part of the body, except directly over the bronchial tubes. Then pass the *pos.*, either with a small sponge or the hand, up and down over the bronchial tubes, while the *neg.* is held on the spine a little below. A tonic treatment may be given, if necessary.

CATARRH, ACUTE.

The *pos.* should be held at the root of the nose while the *neg.* is held between the shoulders, *S.*

C. If chronic, the *neg.* should be held at the root of the nose, while the *pos.* is held between the shoulders, *P. C.* Treat five or ten minutes. A case of long standing may require three or four treatments a week for two or three months. In these cases, a tonic treatment should be given occasionally, or treatment as in Scrofula, if necessary, which is frequently the case in persons of scrofulous habit.

CANKER IN THE MOUTH.

Cover the Tongue Director with a piece of moist linen or cotton flannel, and attach it to the *pos.* Let the patient hold it on the tongue, and as far back as practicable, while the *neg.* is applied about the jaws and neck and upper part of the body, *P. C.* If the patient is debilitated, give tonic treatment.

CANCER.

Cancer requires both Local and General Treatment, in order to eradicate the disease from the system. The tumor, in all *ordinary* cases, should be treated with the *pos.*, closing the circuit with the *neg.*, as the case may require, bearing in mind that the current flows from one pole to the other and must pass through the diseased part in order to affect it materially. Thus, a cancer in the breast, *pos.* on the diseased part, and *neg.* on the back opposite. A tumor on the side of the neck, *pos.* on the tumor and *neg.* in the opposite hand, etc. It is advisable that the *neg.* should be at the extremities whenever it is practicable. This treatment may be given daily for two or three weeks, as the case may require; then two or three applications a week will be sufficient. Ten minutes' treatment of the tumor will be sufficient when given thus frequently. A general treatment should always be given in connection with the local, by sponging the entire person, except the tumor, with the *neg.*, while the *pos.* is either in the mouth or at the base of the spine, or by applying the *pos.* generally while the feet and the *neg.* are in water. The last (that by placing the feet in water, etc.,) may be given every third time, *P. C.* In malignant cases, a compound battery is sometimes necessary. In these cases, advice of an experienced Electrician is very desirable. If the patient is debilitated, give treatment once a week, by seating on the *pos.* and sponging the body with the *neg.*, *S. C.*, omitting all other treatment, and *avoiding to pass the sponge over the cancer or tumor.*

CHANGE OF LIFE.

The diseases and derangements of the system which frequently manifest themselves at this critical period, are usually reached by electrical treatment, but they are so various it would be difficult to give directions for their treatment here; the instructions would be too extended for a work of this kind. *See Flooding.*

CIRCULATION OF THE BLOOD.

A thorough sponging with the *neg.* while the patient is seated on the *pos.*, or with the *pos.* on the tongue—if there is a tendency of blood to the head—will equalize the circulation, *S. C.*

CORNS.

Fasten a small sponge on the point of the Ear Insulator and hold it directly on the corn, (it being attached to the *pos.*,) then close the circuit with the *neg.* on some other part of the foot, *P. C.*

CHILBLAINS.

To avoid them keep the feet warm and dry, and wear loose shoes. Apply the *pos.* to the disease and the *neg.* to some other part of the foot. Slippers of wash-leather worn next to the feet, will aid in effecting a cure, and are considered by many to be a preventive of the disease.

CRICK IN THE NECK.

Pos. in the mouth and *neg.* on the crick, *S. C.*

CRICK IN THE BACK.

If the patient is strong and robust, or if there is a tendency of blood to the head, apply the *neg.* to the crick while the *pos.* is held in the mouth; otherwise, the patient may be seated on the *pos.*, *S. C.*

CONSUMPTION.

Many cases which have been *supposed* to be consumption in an *advanced* stage, have been cured by electrical treatment; but probably those who had charge of these cases were mistaken, either in the disease or as to the condition of the patient. But, without doubt, many cases of *real Phthisis Pulmonalis*, or consumption of the lungs, may be cured in the *early* stages by electrical treatment and a proper mode of living.

Treatment.—Sponge the lungs with the *neg.* while the *pos.* is held on the back, opposite or a

little above the *neg.*, passing from the upper
part of the lungs to the diaphram with the
neg., and from the base of the brain to between
the shoulders with the *pos.*, keeping, as before
directed, the latter opposite, or—what is still
better—a little above the former, *S. C.* Should
there be an unnatural sensitiveness about any
part of the lungs or bronchial tubes on the application of the current, finish the treatment by
reversing the poles and changing the current,
holding the *pos.* for a moment on the sensitive
parts, while the *neg.* is on the back, using *P. C.*
If there is much irritation of the throat the
pos. may sometimes be used in the mouth, connected with the Throat Director, instead of on
the back, while the *neg.* is applied to the lungs.
The above treatment may be given from eight
to ten minutes daily, for a week or ten days,
and afterwards less frequently. A Tonic treatment should also be given four or five times a
week. *If accompanied with hemorrhage the
neg. must not be applied to the lungs.*

It should be remembered that the lungs, in a
normal condition, are more sensitive to the action
of Electricity than some other parts of the
body, hence, we employ a light current in treating them.

3

Regular and daily exercises, which are calcu-
lated to equalize the circulation, expand the
muscles of the chest and inflate the lungs, will
aid in effecting a cure.

The patient should also live mostly out of
doors; not shutting himself up in the house
entirely, even on a stormy day; for by so doing,
he will increase his sensitiveness to the changes
in the weather and his liability to take cold, and
in this climate, during some seasons of the year,
might remain in-doors most of the time if he
ventured out only in fine weather. It is much
more difficult for women to do this than men,
on account of their habits and style of dress;
but it *can* and (if they would recover their
health) *should* be done. She must adopt a style
of dress adapted to this mode of living; one
that will protect her from dampness and cold,
and at the same time be as light as possible;
and one, also, that will allow a free use of the
muscles of the body. It is a great mistake to
dress the body so much warmer than the extremities. The latter are so much farther from
the heating apparatus of the system than the
former, that it really requires an extra amount
of clothing to keep up the equilibrium of heat
in them. Loose, warm shoes should be worn.

and not exchanged when in-doors for light, thin slippers, as the temperature is much lower, in many apartments, in the lower part of them than elsewhere, and not unfrequently there are cold currents of air passing over the floor from under the doors and windows. The arms, legs and feet should be protected by flannels, wearing *two* or *three pairs*, if necessary, to keep warm. The skirts should be short enough to escape the drabble of a dewy morning, or a damp day, so that there will not be a stock of wet dry goods to dry off around the ankles whenever one sits down to rest. And last, though not least, the weight of the clothing should rest on the shoulders, every band and seam of which should be loose enough to allow the free use of every muscle. The shortening of the skirts also lessens the weight of the clothing and saves the friction (which is not inconsiderable) of sweeping ones, besides leaving the limbs untrammelled and thus rendering it an easier matter to walk, all of which are matters of importance to a feeble woman who can scarcely carry her own weight, and yet, in order to regain her health, must have exercise out of doors, where she may drink in life from an untainted atmosphere and gather strength from sunbeams undimmed by shade or drapery.

COLD FEET.

Seat on the *pos.* and rub the limbs thoroughly with the *neg.;* or, if there is a sense of fullness in the head, with a flushed face, or headache, put the *neg.* in the sponge and place it under the feet, then sponge the entire person with the *pos.*, *S. C.*

COLDS.

If attended with chilliness, sore-throat, or headache, connect the throat instrument with the *pos.* pole and let the patient hold it in his mouth, as far back on the tongue as practicable, closing the mouth on it in order to make contacts with the roof of it and the tongue, and sponge the lungs and chest with the *neg.*, using a light current. Wipe dry and rub with the hand until warm, then sponge the arms and spine in the same manner, using a stronger current and rubbing vigorously. If the mouth is too sensitive to admit of a strong current, treat only the upper half of the body while the *pos.* is in the mouth, and the lower half while it is at the base of the spine. If there is chilliness without sore-throat or headache, the entire treatment may be given with the *pos.* at the base of the spine.

If, instead of chills, there is fever, put the *neg.* in water with the feet and give a thorough sponging of the entire person with the *pos. S. C.* It may also be necessary to treat the throat. The patient should be kept warm during the treatment, and a warm foot-bath should be given during the whole of it.

Colds are the fruitful causes of a large proportion of all the diseases which flesh is heir to, hence the importance of breaking them up immediately, which can be done effectually by one or two thorough electrical treatments.

COLD, IN THE HEAD.

Treat as acute catarrh. One or two treatments of ten minutes is usually sufficient to reach these cases if taken immediately, otherwise it may require several treatments. Use a light current on the head, *P. C.*, with the regulator partly in the helix, if not too strong to be agreeable, or a light *S. C.*, if the patient can bear it. The treatment may be repeated in six or eight hours, unless the patient is decidedly better. If attended with chills, treatment may also be given with the *pos.* in the mouth while the *neg.* is passed up and down the spine and over the stomach, liver and bowels, at the same time give a warm foot-bath.

CONTRACTED MUSCLES.

Apply the *neg.* to the contracted muscle, and close the circuit with the *pos.* as the case may require. If muscles of the arms are contracted place the *pos.* in the hand, or between the shoulders; if those of legs, place the *pos.* at the feet or at the base of the spine, etc., *S. C.* mostly, though not entirely.

CROUP.

Hold the *pos.* on the throat in front, (not on the lungs), for four or five minutes, while the *neg.* is passed *up and down* the spine from the shoulders to the base, then hold the *pos.*, with a small sponge wrapped around it, at the root of the nose for three or four minutes while the *neg.* is held between the shoulders; after which, seat the patient on the *neg.* and sponge the neck and shoulders with the *pos.*, bringing it up about the ears also. The second treatment should follow the first immediately, and both may be repeated in an hour if the patient is not relieved. Should there be much fever give general treatment as in fever, *P. C.* Take the regulator entirely out when treating the head. Keep the patient warm during treatment. This treatment is almost unfailing.

CRAMPS.

The *neg.* should be applied to the cramping muscles, closing the circuit with the *pos.* as the case may require.

Example.—Cramp in the stomach : apply the *neg.* to the stomach while the *pos.* is held on the tongue or at the base of the brain. Cramp in the legs : rub the legs with the *neg.* while the patient is seated on the *pos.*, *S. C.*

CHOLERA, MORBUS.

Apply the *neg.* to the stomach, bowels and spine, while the *pos.* is at the base of the spine, *S. C.* Care should be taken that the patient does not get chilled while being treated. It is also very important that he should be kept quiet.

CHOLERA, INFANTUM.

Treated as above, with *P. C.*

CHOLERA, ASIATIC.

Treated as in Cholera Morbus. In an advanced stage, however, the *neg.* should be applied not only to the stomach, bowels and spine, but to the entire person, rubbing the limbs with a strong *S.* current. If necessary, the treatment may be repeated in half an hour or longer with *P. C.*, treating the stomach, bowels and spine especially. The treatment should be given in bed, and the patient, as far as possible, kept covered with flannels, with bottles of warm water about the person, and particularly at the feet, as it is very important that he should be kept warm. Keep him quiet also, and as hopeful and cheerful as possible. Fear of cholera may, without doubt, induce the disease, therefore do not talk cholera in his presence. This treatment will not preclude the use of other remedies, but, if given properly and in time, we believe it would seldom be necessary to call in any such aid.

DEAFNESS.

Deafness may result from a disorganization of the structure of the ear, from paralysis of the auditory nerve, from obstructions, etc.

If from disorganization, there is, of course, no cure. If from paralysis, there is hope in the case. Treatment: Fasten a bit of sponge on the ear instrument and connect it with the *neg.* pole, then introduce it carefully into the ear, and while it is held there by yourself, or the patient, apply the *pos.* electrode with a small wet sponge on the back of the head between the centre and the other ear. Treat in this manner from five to fifteen minutes, commencing with a very light current and increasing it gradually. Another polarity would be, to seat the patient on the *pos.* and apply the *neg.* all about the ear, using the same precaution about the strength of the current; use *P. C.* and *S. C.* alter..ately.

If caused by obstructions, the *pos.* may be held in the ear, in the mouth, and at the root of the nose alternately, while you apply the *neg.* to the back of the neck. When in the mouth use the Throat Insulator, *P. C.*

DEBILITY.

Give Tonic treatment; and if there is particular weakness across the chest, sponge that also with t e *neg.* while the *pos.* is held on the back opposite.

3*

DYSPEPSIA.

Apply the *neg.* to the stomach, liver and bowels, (if constipated,) while the *pos.* is held on the tongue, or at the base of the brain. Tonic treatment should also be given.

Should there be pain or soreness in the region of the stomach on passing the electricity through it, instead of the above treatment, apply the *pos.* to the stomach while the *neg.* is held on the spine opposite and a little below, *S. C.* It frequently requires a long course of treatment to effect a cure where the disease has been of years' standing, though good results may be obtained from a few treatments.

DIPHTHERIA.

This disease requires prompt and efficient treatment.

When attended by a hard, full pulse and flushed countenance, give general treatment by placing the feet in warm water with the *neg.* and sponging the entire person with the *pos.* If the pulse is weak, give general treatment by placing the *pos.* at the base of the spine and sponging the body with the *neg.* In all cases, when it is consistent, treat the throat by holding the *pos.* in the mouth, connected with the Throat Director, while the back of the neck, the shoulders, arms, spine and upper part of the body is sponged with the *neg.* If this cannot be done, then apply the *pos.* on the sides of the neck and about the ears, while the patient holds the *neg.* in the opposite hand; and, also, just above the nose while the *neg.* is held between the shoulders. *P. C.* is to be used, except in the Tonic treatment.

DIABETES.

Most cases may be cured in the early stages by Tonic treatment. Use *P. C.* and *S. C.* alternately.

DROPSY, GENERAL.

Sponge the entire person with the *neg.* while the patient is seated on the *pos.*, or, while he holds it in his mouth; occasionally sponging the person with the *pos.* while the *neg.* is in water with the feet, *P. C.* If there is any local disease it should receive especial treatment.

EAR-ACHE.

Pos. in and about the ear while the patient holds the *neg.* in the hand opposite to the affected ear. Commence with a light current and increase gradually. Treat until relieved, *S. C.*

SWELLING OF THE EARS.

Treated as above, only it may not be necessary to use the instrument in the ear unless the internal ear is swollen.

EARS, RUNNING OF THE.

Sometimes the ears discharge at intervals, or constantly for months and even years. Whenever this condition is accompanied with pain or soreness in or about the ear, treat as above. But when this is not the case, and there has been a fetid discharge more or less for some time, treat two days successively by seating on the *pos.* and applying the *neg.* to the entire body, but more especially in and about the affected ear or ears ; changing the treatment every third day by putting the *pos.* on the tongue, instead of at the base of the spine, then treating as above with the *neg.* Should the first method of treatment make the head ache, it may be finished up by wetting the head and applying the *pos.* to it for two or three minutes, while the patient holds the *neg.* in his hands, *P. C.* Give these treatments twice a week until there is a decided improvement, then less frequently. Should there be scrofula in the system, an occasional treatment should be given by putting the feet in water with the *neg.* pole and sponging the *entire* person, limbs and all, with the *pos.* The ears should be kept clean by syringing them out with warm water and a little castile soap.

ENLARGED GLANDS.

The *pos.* should be applied to the affected glands, while the circuit is closed with the *neg.* as the case may require, *P. C.* If necessary, give general treatment as in scrofula.

EPILEPSY.

The condition of the patient during the paroxysm, (or fits,) and during the intervals is quite different and requires very different treatment.

During the paroxysms the treatment is the same as that of apoplexy, except that the head is treated *more* and the spine *less.* The time to cure, however, is during the intervals. A careful examination should be made to ascertain if there is any derangement of the organs which causes the fits. If this is the case, local treatment must be given to remove the cause, and general treatment to give tone to the system, *S. C.* If there is no such cause existing, Tonic treatment is all that is required. Give it by sponging the entire person with the *neg.*, especially the stomach and bowels, while the *pos.* is either at the base of the spine or on the tongue. The two treatments may be given alternately, unless there is a tendency to a rush of blood to the head. In that case treat mostly with the *pos.* on the tongue, *S. C.* It may require two or three treatments a week, for months, to effect a cure, though some cases yield to treatment very readily. Great care should be taken to have the diet light and digestible. All excitement should be avoided.

TREATMENT OF THE EYE.

The usual treatment of this delicate organ by oculists and physicians is severe and tedious, and it may be satisfactory to those who wish to try Electricity, to know that there is no pain or unpleasant sensation attending this mode of treatment when properly administered. If the eye has been diseased long, it will take *time* to cure. But, if *curable*, it is a safe and sure method of accomplishing it, if persevered in. Severe cases of *long* standing of inflammation, where the eye is ulcerated, or the lids granu-lated and thickened, may require an occasional treatment for a year or more. (This is more particularly the case when the eye has been treated a good deal with Costics, Blue Stone, etc.) Acute inflammation of the eye yields to electrical treatment very readily; from one to three applications usually effects a cure.

In all these cases great care should be taken to keep up a good circulation. Keep the hands and feet warm, and, if the patient is confined in a dark room, let him have a change of rooms, so that one can be thrown open to the *sun* and air while he occupies the other. The stomach and bowels should also be kept in good order, —not by drugs,—but by proper diet and elec-trical treatment. Enemas of water may be given, if necessary.

EYES, INFLAMMATION OF.

Attach the eye-glass to the *pos.* pole and fill it with *pure soft* water, then hold the eye in it while the *neg.* is held in the hand on the corres-ponding side. The handle of the glass should be kept perpendicular, and the head brought

forward in a position that the eye may be en
tirely immersed in the water. The water should
be changed once or twice during the treatment.
Should there be pain about the eye, or in the
head, use the *pos.* on the seat of the pain while
the *neg.* is held in both hands, or, if the feet are
cold, place it in water with the feet. Treat
the eye from ten to thirty minutes as the case
may require. Use *S. C.* Should the inflammation be caused by a derangement of the system, general or local treatment must also be
given, according to the condition of the case.
A general treatment with the *neg.*, while the
pos. is on the tongue, is also advisable.
*Avoid rubbing the eyes with the hands in all
cases.*

GRANULATED, OR THICKENED LIDS.

The treatment of the eye should be the same
as above, except that the *P. C.* should be used,
and that usually general, and often other local
treatment is necessary. A general treatment
with the *neg.* while the *pos.* is held on the
tongue is good in most cases. If there is scrofula in the system, which is frequently the case,
treat accordingly. If there is much debility,
occasionally seat the patient on the *pos.* and

sponge the entire person with the *neg.*, using *S. C.*

ULCERATION OF THE EYE.

Treat precisely as the above.

OBSTRUCTION OF THE LACHRYMAL DUCT.

Pos. on the disease, *neg.* on the back of the neck, or in the opposite hand. *P. C.*

BLINDNESS FROM OPACITY OR FILM ON THE EYE.

Pos. at the eye, as in inflammation,—*neg.* in the hand. *P. C.* An occasional treatment about the eye with the *neg.* for three or five minutes, while the *pos.* is on the spine, will promote absorption and hasten a cure. *S. C.* should be used in this part of the treatment. Most cases of blindness from this condition can be cured by persevering treatment.

CATARACT.

The *pos.* should be used at the eye with eye-glass—as in inflammation of the eye—and the *neg.* at the base of the spine. *P. C.* Give general treatment if necessary, and as the case may require. But few cases can be cured by an or-

dinary Electrical Machine, as they usually re:
quire a compound battery. The advice of an
experienced Electrician is desirable in these
cases. When curable they require a long course
of treatment. In most cases give Tonic treat- .
ment occasionally.

AMAUROSIS.

The treatment for the eye is *neg.* at the eye
with eye-glass filled with tepid water, and *pos.*
at the base of the spine. Should there be
much pain in the back of the head, the *pos.*
may be held there a portion of the time instead
of at the base of the spine. *S. C.* If the patient is debilitated, or if there is much derangement of the nervous system, a Tonic treatment
should be given two or three times a week
while the treatment of the eye may be given
from four to six times a week until there is
a decided improvement in the case, and then
less frequently. The case should be diagnosed
carefully, as there is frequently some derangement in the system which causes, or, at least,
aggravates the disease, and which must be removed in order to effect a permanent cure. A
large portion of cases, however, are curable,
though it may require considerable treatment.

FACE-ACHE.

If attended with heat and redness of the face, apply the *pos.* to the face, while the *neg.* is either at the feet or at the base of the spine. If attended with paleness of the face and debility, give Tonic treatment.

FELONS.

Place the hand or finger affected in a small vessel of water with the *pos.* while the *neg.* is held in the opposite hand. *P.C.* The suffering attending this painful disease may be greatly relieved at any stage, but to prevent suppuration it must be taken at an early stage. If very painful, a treatment (of ten or fifteen minutes) may be repeated every two or three hours.

FEVERS.

It would be a difficult task, in a work of this kind, or, indeed, in any work, to give instructions to the inexperienced to treat fevers in their various types and stages. Yet most attacks may be broken up, or, at least, greatly modified, if taken in an early stage.

Treatment.—In sudden attacks, (or whenever the fever is high,) put the feet in warm water with the *neg.* and sponge the entire person with the *pos. N. C.* One or two treatments, with rest and abstinence from food for a day or two, is usually all that is necessary in sudden attacks. Should there be any local difficulty, give treatment accordingly. And, if necessary, follow by Tonic treatment.

Typhoid fever, if not complicated, will be thrown off by Tonic treatment in most, if not in *all* cases. And should it be complicated, or should there be local derangements, give especial treatment as the case may require. It is *particularly* important in this type of fever that the patient should avoid all mental and physical effort. He should be kept quiet, but cheerful, and free from excitement. Great care should be taken in reference to diet, even when convalescent.

ERUPTIVE FEVER.

If the eruption comes to the surface properly, treat as in the *first* case of fever, except that the *P. C.* should be used. If it does not, first give general treatment with *neg.* Place the *pos.* in the mouth, and sponge the upper portion of the person with the *neg.*; then place it at the base of the spine, and sponge the lower portion

limbs and all.   This is to bring the eruption to
the surface.   After which, treat as first directed,
by placing the feet in warm water with the *neg.*,
and sponging the entire person with the *pos.*
If the child is too young to hold the *pos.* in the
mouth, that part of the treatment must of
course be omitted, although it is very important.   If the throat is sore, or if there are
swellings about the ears or throat, apply the
pos. thoroughly about the ears and neck while
the patient holds the *neg.* in the hand opposite
to the side which is being treated.   *P. C.* should
be used in the entire treatment.   Great care
should be taken to ventilate the sick room
thoroughly in these cases.   The above is the
treatment for Measles, Scarlet Fever, etc.   Keep
the patient warm during treatment.

FEVER AND AGUE.

Apply the *neg.* over the liver while the *pos.*
is held on the spine opposite, or in the mouth.
This, with thorough tonic treatment, which
should be given by seating the patient on
the *pos.* and sponging the body, but more especially the spine with the *neg.*, is all that is
necessary to break up the most obstinate cases.
A strong secondary current should be used.   A

daily treatment should be given as long as necessary, which will seldom exceed a week.
Treat just before the time for the chill to come
on.

FEVER AND CHILLS.

Treat as in Fever and Ague, just before the
time for the chill to come on. A treatment
may be given for the fever if it runs high, but
this will seldom be necessary.

FLOODING, OR MENORRHAGIA.

Connect the Womb Insulator with the *pos.*
pole, and introduce it into the vagina; then apply the *neg.* to the back and upper portion of
the bowels. This treatment should be given
while the patient is in bed. Commence with a
light current, and increase it gradually, until
the patient has as much as she can bear com-
fortably. *S. C.* Avoid shocks from the battery,
also all excitement. A general treatment with
the *neg.*, while the *pos.* is at the base of the
spine, should be alternated with the local treatment.

GOITRE.

Apply the *pos.* to the tumor, while the *neg.*
is held in the patient's hands, on the back of

the neck (below the *pos.*) and at the base of
the spine alternately.   Give a general treat
ment occasionally with the *neg.*, also treat occasionally as in scrofula.   *P. C.* should be used
on the tumor.   A treatment of fifteen or twenty minutes should be given three or four times
a week.   If it is very large it may require sev
eral months to effect a cure.

GRAVEL.

This disease may be cured by proper electrical treatment, but it requires skill and experience.

HOARSENESS.

The *pos.* should be held in the mouth connected with the Throat Insulato ;, and over the
throat in front, while the *neg.* is passed over
the spine. *P. C.* If the hoarseness has become
chronic and is accompanied with debility, give
Tonic treatment.

HEART.

Diseases of the heart are treated very suc-
cessfully with Electricity, but, in complicated
cases, they require skill and experience.   They
may also require a long course of treatment, as

the patient is frequently so sensitive that the lightest current must be employed, and a very short treatment given, especially when unaccustomed to the sensation produced by Electricity, and when there is inflammation.

If there is an enlargement of the organ, or a thickening of the valves, or any obstruction, or if there is inflammation or dropsy, the *pos.* pole should be applied over the organ while the *neg.* is held on the back a little below the *pos.* and a little to the right side or at the base of the spine; the two polarities may be alternated. In these cases the patient is usually very sensitive to the current at first, so that it is sometimes necessary to reduce the strength of the solution in using the primary current, which is the current to be used in all of these cases. Be careful that the patient receives no sudden shock from the current. If there is general Dropsy, give general treatment with the *pos.* while the *neg.* is at the feet.

If it is merely a nervous affection, Tonic treatments will usually prove effective. Commence with a light current, increasing it gradually according to the sensitiveness of the patient.

HEADACHE. .

If attended with redness of the face, heat in the head or flow of blood to it, wet the head thoroughly to the scalp and apply the *pos.* carefully to it, while the *neg.* is at the base of the spine or at the feet. Use a very light secondary current, if the patient can bear it, otherwise *P. C. Never treat the head with a current strong enough to cause severe pain.* Should the patient be chilly, treat the spine, stomach and liver with the *neg.* while the *pos.* is held on the tongue. When attended with paleness of the face, nervous prostration, or general debility, give Tonic treatment.

SICK HEADACHE.

Pos. on the base of the brain and *neg.* over the stomach, liver, (and bowels, if constipated.) This treatment will reach most cases; still there are different causes for sick headache, and in order to effect a cure the cause must be removed. Constipation of the bowels is one of the most fruitful causes. (See Constipation.)

HERNIA.

Pos. on the disease and *neg.* on the back opposite *S. C.*

4

HIP DISEASE.

Pos. on the hip and *neg.* at the foot—*P.* and *C.* current alternately. Give Tonic treatment also, if the patient is debilitated, or treatment as in Scrofula, if necessary, which is frequently the case.

HYSTERIA.

Pos. on the tongue, or on the cerebellum, and *neg.* over the chest, stomach and bowels. *S. C.* Also give Tonic treatment.

INFLAMMATION.

Inflammations are treated with the *pos.*, closing the circuit with the *neg.* as the case may require. *S. C.* usually. *Inflammation of the lungs is an exception to the above rule.*

INSANITY.

Many cases of insanity may be cured by proper electrical treatment, but it requires skill and experience to diagnose the case and decide upon the proper course to be pursued.

INDIGESTION.

Neg. on the stomach, *pos.* passed over the spine from the cerebellum to the base, or on the tongue. The two treatments may be alternated. *S. C.*

INDURATIONS.

The positive pole and primary current are always used in indurations.

KIDNEY, INFLAMMATION OF.

Pos. over the kidneys and *neg.* in water with the feet. If attended with fever, the entire person may also be sponged with the *pos.* while the *neg.* is at the feet *S. C.*

KIDNEY, BRIGHT'S DISEASE OF THE.

First give a few treatments as above. Afterwards treat the entire person (except the kidneys), first, with the *neg.* while the *pos.* is in the mouth for the upper part of the body, and at the base of the spine for the lower half. Then put the *neg.* at the feet as before, and treat the entire person with the *pos.*, but especially the kidneys.

The two last-named applications should be made at one sitting. The *P. C.* is to be used entirely in this disease. This treatment has been followed with the most satisfactory results in cases which had been given up as hopeless by medical attendants.

LEUCORRHŒA, OR WHITES.

In the early stages, introduce the largest silver instrument, connected with the *pos.*, into the vagina, and apply the *neg.* over the bowels and across the hips. If it has been of long standing and the discharge is yellow, brown or greenish, reverse the poles. *P. C.* A Tonic treatment should be given in connection with the local one. Enemas of tepid water with a little Castile soap, or, what is still better, thirty drops of the tincture of Myrrh to half a pint of water, should be given night and morning.

LIVER, INFLAMMATION OF.

Pos. on the liver while the *neg.* is held on the opposite side between the spine and left hip, or at the base of the spine, or under the left foot. These *three* polarities may be used in alternations. Give general treatment as in fever if necessary. *S. C.*

LIVER, TORPID.

Pos. on the tongue, or on the cerebellum, and *neg.* over the liver. *S. C.* when used on the cerebellum, and *P. C.* when used on the tongue. General treatment should also be given.

LIVER, ENLARGEMENT OF.

Treat as in inflammation, using the *P. C.* instead of the *S. C.* Also give longer treatments, as it will require a much longer time to cure the latter than the former.

LOCKJAW.

Neg. just in front of the ear over the contracted muscle, *pos.* in the opposite hand, or on the back of the neck. *S. C.*

LUNGS, INFLAMMATION OF.

Sponge the lungs with the *neg.* while the *pos.* is held on the back nearly opposite, or a little above the *neg.*, passing from the upper portion of the lungs down to the diaphragm. Should there be tenderness about the lungs or bronchial tubes on application of the current, finish the treatment by reversing the poles, holding the *pos.* for a moment on the sensitive parts while the *neg.* is on the back; *S. C.* if not too sensitive. It should be remembered that the lungs, when in a normal condition, are more sensitive to the action of Electricity than some other parts of the body, hence, we employ a light current in treating them.   General treatment, as in fever, will be necessary in most cases.

LOSS OF VOICE, OR APHONIA.

Pos. on the cerebellum, *neg.* over the larynx. *S. C.* This treatment is almost unfailing when caused by paralysis.   Should the patient be debilitated, give Tonic treatment also.

MEASLES.

Treat as in eruptive fevers.

MENSES, SUPPRESSION OF.

Pos. on the tongue and *neg.* over the lower part of the body, but especially over the abdomen.   If the case is obstinate, or one that requires *immediate* relief, introduce the instrument, connected with the *neg.*, into the vagina, and sponge the back and bowels with the *pos. S. C.* strong.

MENSES, PAINFUL.

Treat as above, or with the *pos.* over the bowels while the feet are in warm water with the *neg. S. C.*

MENSES, RETARDED.

The same treatment, if the patient is of full habit, has head-ache attended with redness of the face, etc. If pale, thin, or debilitated, give Tonic treatment instead, avoiding all excitement, fatigue, drugs, etc. *S. C.* in both cases.

MENSES, TOO COPIOUS.

Treat as in flooding, unless Tonic treatment proves sufficient.

MISCARRAIGE.

See flooding.

NEURALGIA.

If attended with soreness or swelling of the parts, apply the *pos.* to the disease, otherwise the *neg. S. C.* Give Tonic treatment in the last case if the patient is debilitated.

NOSE-BLEED.

Hold the *pos.* on the nose between the eyes, and the *neg.* on the back of the neck. *S. C.*

NOSE, SWELLING OF.

Apply the *pos.* to the nose while the *neg.* is held alternately on the back of the neck and in the hands. *P. C.*

OVARIES, INFLAMMATION OF.

Put the *neg.* in water with the feet and apply the *pos.* to the disease, or give a sits-bath with the *pos.* in the bath-tub and the *neg.* at the feet. *S. C.*

OVARIAN TUMOR.

Arrange the poles so that the current will pass directly through the tumor, by holding the *pos.* on it and the *neg.* on the back opposite. Every third treatment apply the *pos.* not only to the tumor, but also over the bowels and back, while the *neg.* is with the feet in water. *P. C.* Treat from ten to twenty minutes, repeating the treatment four or five times a week.

OBSTRUCTIONS.

If attended with inflammation, swelling, or induration, hold the *pos.* on the affected part and the *neg.* opposite, or a little below. *P. C.* Otherwise apply the *neg.* to the disease while the *pos.* is held above, or opposite to it. *S. C.*

OZŒNA.

Hold the *pos.* on or over the disease, and the *neg.* opposite or below. If this treatment does not relieve the difficulty in six or ten days, reverse the treatment, holding the *neg.* on the disease and the *pos.* opposite. *P. C.* In the last case give from three to six treatments a week, also general treatment, as in scrofula.

PALSY, OR PARALYSIS.

Neg. over the paralyzed parts, closing the circuit with the *pos.* as the case may require. If one arm is paralyzed, apply the *neg.* to it while the *pos.* is held in the opposite hand, or on the spine. In paralysis of one side, apply the *neg.* to it, and at the same time pass the *pos.* over the opposite side. If the lower half is paralyzed, hold the *pos.* in the hands and on the upper part of the spine alternately, while the *neg.* is applied to the part affected, etc. *S. C.*, with an occasional use of the *P. C.*

In either case, if accompanied with redness of the face, an unnatural flow of blood to the head, or a sense of fullness in the head, give a general treatment of the body with the *neg.*, while the patient holds the *pos* on the tongue. If, instead of the above symptoms, the patient has a feeble pulse, is pale and debilitated, give thorough treatment with the *neg.* while the *pos.* is at the base of the spine.

4*

If there is a flow of blood to the head, accompanied with cold feet, give an occasional treatment by placing the *neg.* at the feet and sponging the entire person with the *pos. S. C.*

PILES.   (SEE ANUS.)

Connect the rectum instrument with the *pos.* pole, and after oiling it well, introduce it carefully into the rectum and apply the *neg.* to the lower part of the spine. *P. C.* If the parts are too much swollen to admit of using the instrument at first, take a small soft sponge and wrap it around the electrode, and press it against the parts, while the *neg.* is applied to the spine. If persevered in, the treatment is unfailing. Avoid cathartics.

PLEURISY.

Treat as in Inflammation of the Lungs.

POLYPUS.

Pos. on, or directly over the tumor; *neg.* as the case may require. If in the nose, *neg.* on the back of the neck. If in the womb, on the back opposite and on the bowels. *P. C.*

PREGNANCY.

Electricity will relieve much of the suffering

and discomfort attending pregnancy, especially in the early stages, if applied properly, and used with care. A slight, even current should be employed, avoiding shocks and sudden contraction of the muscles. The *neg. should never be placed at the base of the spine or abdomen.* Applying the *neg.* to the back, stomach, bowels and limbs, while the patient is seated on the *pos.*, is quieting to the nervous system, promoting digestion, and inducing sleep, and relieving many of the discomforts attending the condition. It is also a preventive to miscarriage. *S. C.*

QUINSY AND SORE THROAT.

Fasten a soft sponge, or a piece of cotton flannel, over the ball of the throat instrument, and connect it with the *pos.* Place it as far back in the mouth as comfortable, closing the mouth on it in order that it may make contact with the tongue and roof of the mouth, then sponge the shoulders, arms and spine thoroughly with the *neg. P. C.*—commence with a light current, increasing it gradually. If feverish, give general treatment also, as in fever, applying the *pos.* thoroughly about the ears and throat, as well as over the entire person, while the *neg.* and feet are in the warm water. In any case of sore throat, if the patient is too young, or, from any other cause, cannot hold the instrument in the mouth, the *pos.* may be applied externally round the ear and throat, treating one side at a time, while the patient holds the *neg.* in the hand opposite to the side being treated.   Give treatment also as in fevers.

RHEUMATISM.

Acute or Inflammatory Rheumatism is treated with the *pos.* If about the head, in the neck, shoulders, or arms, the *neg.* should be placed in a basin of water with the hand or hands, while the *pos.* is applied to the painful or swollen parts. If it is located in other portions of the body the *neg.* may be placed at the feet. Most cases require a general treatment, as in fevers : sponging the entire person with the *pos.*, while the *neg.* is in water with the feet. If there is much swelling, or enlargement of the joints, the primary current should be mostly used in the local treatment, otherwise, alternate the two currents.   In severe cases, which are attended with much inflammation or fever, the treatment may be repeated two or three times during twenty-four hours.

RHEUMATISM, CHRONIC.

If it has been of long standing, and is attended with contraction of the muscles, or stiffening of the joints, it should be treated with the *neg.* pole, using *P.* and *S. C.* alternate. Should this treatment continue to aggravate the symptoms after two or three applications, an occasional treatment may be given by reversing the poles. It requires a long course of treatment to cure cases of this kind, of years' standing, but the sufferings are often greatly relieved by a few applications, though the first treatments frequently aggravate them, which should not be considered an unfavorable symptom. If there is an enlargement of the joints, it will be necessary to treat them, at first, with the *pos.* pole and *P. C.* In both cases Tonic treatment should be given.

SCROFULA.

Sponge the entire person with the *neg.* while the *pos.* is either in the mouth or at the base of the spine. If practicable, the *pos.* should be held in the mouth, connected with the throat director, while the *neg.* is applied to the upper portion of the body, then placed at the base of the spine, while the *neg.* is applied to the lower part of the body and lower limbs.

Every third treatment should be with the *pos.* over the entire person, while the *neg.* is in a vessel of warm water with the feet.

The head should not be treated unless affected. Treatments given from three to six times per week.

If there are tumors, enlarged glands or ulcers, give treatment according to directions given in these cases. *P. C.* should be usually employed; a tonic treatment being given occasionally with the *S. C.*

Fresh air, sunlight, plain, nutritious diet, and out-door exercise, as far as the patient is able to take it, are indispensable in effecting a cure.

SORE MOUTH.

Cover the Tongue director with moist linen, and connect it with the *pos.* pole, then let the patient hold it on the tongue, while the *neg.* is applied to the lower part of the face, the neck, shoulders, spine, and stomach. Use *P. C.,* and commence with it very light, increasing the strength as the patient becomes accustomed to the sensation. If attended with debility, give Tonic treatment.

SORE NIPPLES.

Connect the eye-glass with the *pos.* pole and fill it with tepid water, then hold the nipple in the water, while the *neg.* is held in the hand on the corresponding side. Treat in this manner five or ten minutes. *S. C.*

SLEEPLESSNESS.

If the patient is naturally feeble, or is debilitated from any cause, seat on the *pos.* and sponge the entire person with the *neg.*, using a light *S. C.* This treatment is frequently given with good results just before retiring. If, however, the patient is restless from fever, or any exciting cause, put the feet in warm water with the *neg.*, and sponge the entire person with the *pos.*, or sponge with the *neg.* while the *pos.* is held in the mouth.

SPASMS.

Should usually be treated with the *neg.*, giving such other treatment to remove the cause as the case may require.

SPINE, CURVATURE OF.

Always use the *neg.* on the contracted muscles

and the *pos.* on the expanded ones, closing the circuit with the other pole by placing it either in the mouth, in the hands, or at the base of the spine. If the curvature is in the upper part of the spine, the *best* polarity is from the mouth, treating first the contracted muscles, then reversing the poles for the expanded ones, occasionally changing the treatment by sponging the chest with the *neg.* while the *pos.* is held on the expanded muscles; but never use the *pos.* on the chest while the *neg.* is on the spine. If the curvature is in the lower part of the spine, the best polarity is from the base of the spine. A Tonic treatment should be given occasionally in all cases. In young persons the treatment is very successful, and in older ones will relieve the suffering and arrest the disease. *P.* and *S. C.* alternate. The throat instrument should be used in treating from the mouth.

SPINE, INFLAMMATION OF.

Place the feet in water with the *neg.* and sponge the spine with the *pos. P.* and *S. C.* as above.

SPLEEN, INFLAMMATION OR ENLARGEMENT OF.

Apply the *pos.* to the diseased part and the

neg. on the back opposite, or at the feet. *P. C* in enlargement, and *S. C.* in inflammation.

STAMMERING.

Many cases of stammering may be cured by applying the *neg.* over the throat and chest, while the *pos.* is held at the base of the brain. If the child is feeble, seat on the *pos.* and apply the *neg.* also to the entire spine. *S. C.*

STOMACH, INFLAMMATION OF.

Pos. on the stomach, *neg.* on the spine a little below. *S. C.*

ST. VITUS' DANCE.

Wet the head thoroughly; apply the *pos.* to the cerebellum (small brain) and *neg.* to cerebrum, (large brain) from three to five minutes at a time, repeating the treatment from two to four times a week. *P. C.* so light that it will scarcely be perceptible to the patient.

A Tonic treatment should be given two or three times a week. *Pos.* on the tongue, and *neg.* over the body, is also a good treatment in many cases. *S. C.*

SUNSTROKE.

If there is redness of the face and headache,

put the feet in warm water with the *neg.* Wet
the head thoroughly and apply the *pos.* *P. C.*
very light.

TONIC TREATMENT.

The more general way of giving Tonic treatment is by placing the *pos.* pole at the base of
the spine—by seating the patient on it, or otherwise—and sponging the entire person (head
excepted) with the *neg.* But, if in case of debility, there is also a tendency to sore throat or flow
of blood to the head, sore eyes, or eruption of the
face, the *pos.* pole may be attached to the throat
or tongue instrument, and held in the mouth by
the patient, while the upper half of the body is
sponged with the *neg.*, then placed at the base
of the spine while the lower half is sponged
with the *neg.* *S. C.* If there is a weakness of
the lungs, chest, or digestive organs, or in any
case where there is great prostration of the vital
forces, commence the treatment by applying the
neg. over these organs, while the *pos.* is held on
the spine a little above the *neg.*, then follow
with the above treatment. *S. C.* [*See General
Treatment for further directions, Chap. III.*]

TOOTHACHE.

The *pos.* should be applied to the tooth and

face, while the *neg.* is held in the patient's hand,
S. C. If ulcerated, the primary current should
be used, and a prolonged treatment given. Two
or three treatments of twenty minutes will not
only cure, but, usually, break up the tendency to
ulceration.

TUMORS.

Apply the *pos.* to the disease, closing the circuit with the *neg.*, as the case may require. If
about the neck or head, the *neg.* may be held in
the hands ; if on the body, it may be necessary
to hold the *neg.* on the opposite side, as the current must pass through the tumor to effect it materially ; if on the lower limbs, it may be placed
at the feet. *P. C.* General treatment should be
given as in Scrofula.

URINE, RETENTION OF.

Neg. over the bladder, *pos.* on the back opposite. If there is difficulty in retaining the urine,
seat in the *pos.* and pass the *neg.* over the lower
part of the bowels. *S. C.*

UTERUS, PROLAPSUS OF—(OR FALLING OF THE WOMB.)

Connect the instrument with the *pos.* and in

troduce it into the vagina far enough, if possible, to make contact with the mouth of the uterus, then apply the *neg.* to the back, as far up as the shoulders, and over the upper part of the bowels, from ten to twenty minutes. If practicable, use the Uterine Insulator, otherwise the largest silver instrument. If the disease has been of long standing, this treatment should be alternated with Tonic treatment, and given daily for ten or twelve days, and afterwards less frequently. Should it arouse the system too suddenly, give a light current, otherwise, a good *S. C.* The patient need not be alarmed, however, if the symptoms are aggravated at first, as this is frequently the case when the *best* results are being obtained, especially in cases of long standing. It is advisable that the patient should keep a horizontal position for an hour or two after treatment. A Tonic treatment should be given three or four times a week in all severe cases.

If persevered in, the above treatment is almost *unfailing,* even in cases of years' standing. But it requires time, and an observance of some of the general laws of health. The patient should, if possible, take a little gentle exercise out of doors, or, if she can do no better, allow herself to be carried out in her chair and set in the sun for a little while, each day, until able to walk. Her diet should be plain and nutritious. Her apartments well sunned and ventilated, and, when able to walk, the utmost care should be taken to have her clothing loose and light, and supported by her shoulders, in order that there may be no pressure or weight on the relaxed muscles. Indeed, it would be advisable for all ladies to observe these rules.

UTERUS, INFLAMMATION OF.

Apply the *pos.* over the uterus, or, give a warm sitz-bath with the *pos.* in it, while the *neg.* is at the feet. If attended with fever, apply the *pos.* to the entire person, while the *neg.* is at the feet *S. C.*

UTERUS, ULCERATION OF.

Use the Womb Insulator attached to the *pos.*, as in prolapsus, and apply the *neg.* over the bowels and back for ten minutes, after which, apply the *pos.* over the lower part of the bowels while the *neg.* is held nearly at the base of the spine for the same length of time. An occasional treatment of the bowels with the *pos.*, while the *neg.* is at the feet, is advisable; also Tonic treatments. *P. C.*, except in the Tonic treatment.

An enema of thirty drops of the Tincture of Myrrh to a half pint of tepid water, should be · given twice a day. For further instructions, see Prolapsus Uteri.

UTERUS, ENLARGEMENT AND INDURATION OF.

Treated precisely as ulceration. Both of these cases usually require a long course of treatment to effect a cure, if of long standing; if not, they yield readily.

VAGINA, INFLAMMATION OF.

Connect the instrument with the *pos.* and introduce it into the vagina, then apply the *neg.* with a wet sponge over the lower part of the back and over the abdomen. *S. C.*

VOMITING.

Hold the *pos.* on the cerebellum (or base of the brain), while the *neg.* is passed over the stomach and liver, and bowels, also, if constipated ; the *pos.* may be moved down on the spine if the treatment is continued more than ten minutes, which may be necessary in an obstinate case. *S. C.*

WATER BRASH.

The treatment may be alternated with the

above by holding the *pos.* on the tongue or passing it up and down the spine, while the *neg.* is held on the stomach. *S. C.* Most cases require Tonic treatment.

WHITLOWS ON THE FINGERS.

Pos. on the disçase, or in a bowl of water with it, while the *neg.* is held in the opposite hand. *P. C.*

WHITE SWELLING.

The *pos.* should be applied to the swelling while the *neg.* is at the foot, if the swelling is on the knee or ankle. *P. C.* General treatment will also be necessary, as in Scrofula. If it has been discharging long, it will not require much local treatment; in this case a Tonic treatment, occasionally, would be advisable in connection with the general treatment, as in Scrofula.

WOUNDS.

When inflamed apply the *pos.* while the *neg.* is kept at some point below.

WEAKNESS OF SIGHT.

If attended with general debility, or weakness

of the spine, great benefit may be derived from Tonic treatment.

WEAK STOMACH.

Apply the *neg.* over the stomach, while the *pos.* is held on the tongue, or passed up and down the spine from the cerebellum to opposite the stomach. *S. C.* Tonic treatment should also be given occasionally.

WEAK LUNGS.

Sponge the lungs thoroughly with the *neg.* while the *pos.* is held on the back, commencing at the base of the brain with the *pos.* and on the upper portion of the lungs with the *neg.*, and so pass down, keeping the *pos.* a little above the *neg.* A Tonic treatment would also be advisable. Commence with a light current over the lungs, and increase it gradually. Do not, in any case, give it strong enough to produce pain. *S. C.*

WORMS.

Apply the *neg.* over the stomach and bowels, while the *pos.* is held on the spine. Use the full strength of the primary current. If the child is delicate, give Tonic treatment; or, if necessary, give treatment as in fever.

N O T I C E.

—:o:—

ELECTRO-MEDICAL
INSTRUMENTS & APPARATUS,

FOR SALE BY

S. M. WELLS, Medical Electrician,

LA CROSSE, - · - · WISCONSIN.

———◆•◆———

WELLS' BATTERY, with a full case of Treating
   Instruments, - - - - -
Battery, without a Case of Treating Instruments,
Full Case of Treating Instruments, - -

www.ingramcontent.com/pod-product-compliance
Lightning Source LLC
Chambersburg PA
CBHW021409090426
42742CB00009B/1071